FLOWERS

Ryan Smith and
Heather Kissock

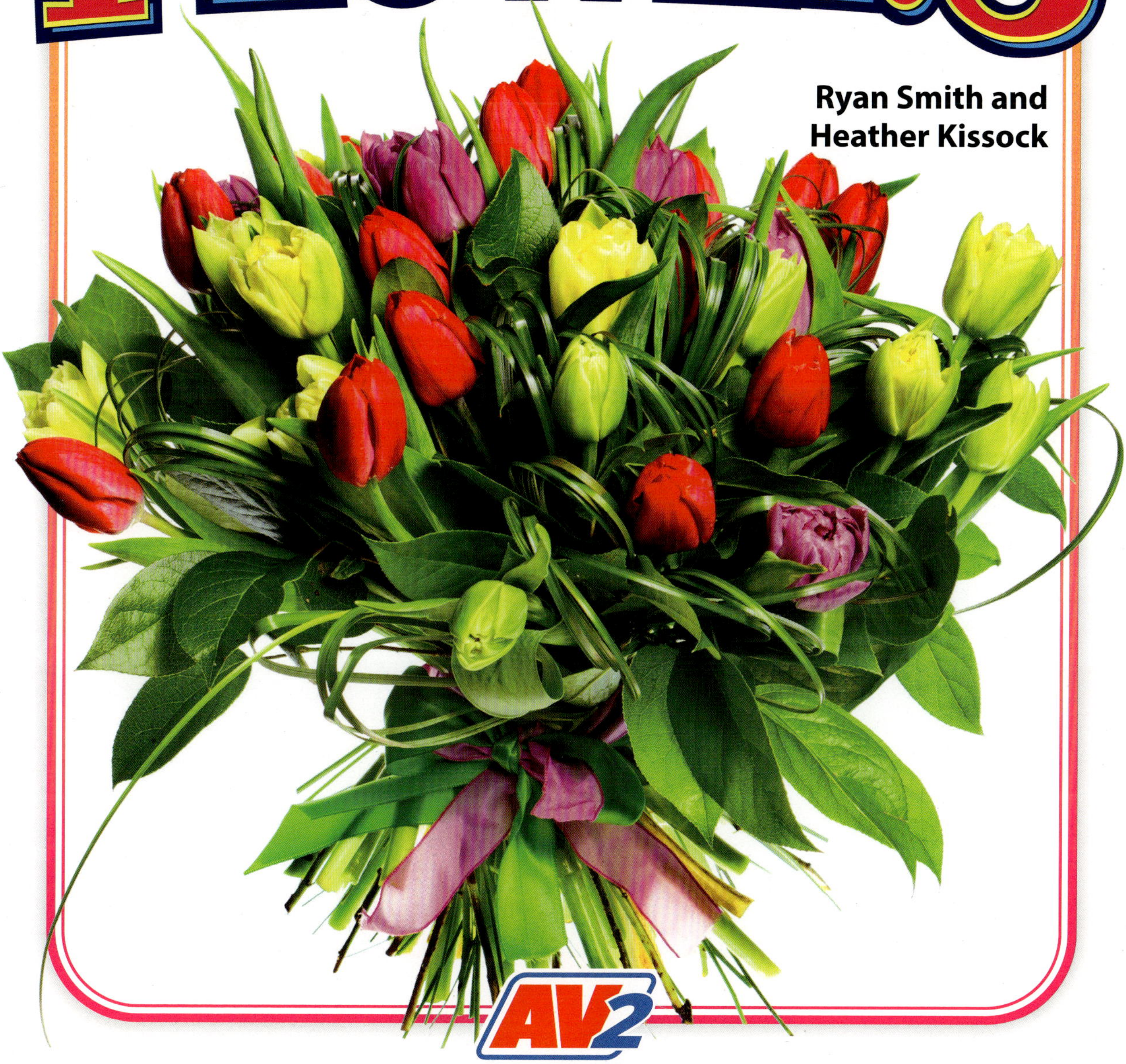

AV2

www.av2books.com

Step 1
Go to **www.av2books.com**

Step 2
Enter this unique code
SBIZGM4GV

Step 3
Explore your interactive eBook!

AV2 is optimized for use on any device

Your interactive eBook comes with...

Contents
Browse a live contents page to easily navigate through resources

Audio
Listen to sections of the book read aloud

Videos
Watch informative video clips

Weblinks
Gain additional information for research

Try This!
Complete activities and hands-on experiments

Key Words
Study vocabulary, and complete a matching word activity

Quizzes
Test your knowledge

Slideshows
View images and captions

This title is part of our AV2 digital subscription

1-Year K–5 Subscription
ISBN 978-1-7911-3320-7

Access hundreds of AV2 titles with our digital subscription.
Sign up for a FREE trial at **www.av2books.com/trial**

CONTENTS

All about Flowers

If you were to grow a flower garden with your mom or dad, what types of flowers would you put in it? Why did you choose these ones? Are they your favorite color? Do they smell nice?

Growing a flower garden is a fun way to learn about plants. You will find out what your flowers need to grow and be healthy. You can take care of them every day. Doing this brings many rewards. The colors will brighten your day!

Gardening can help people learn more about the environment, especially the importance of having clean water and healthy soil.

What Is a Flower?

A flower is the **bloom** of a flowering plant. Flowers have several parts.

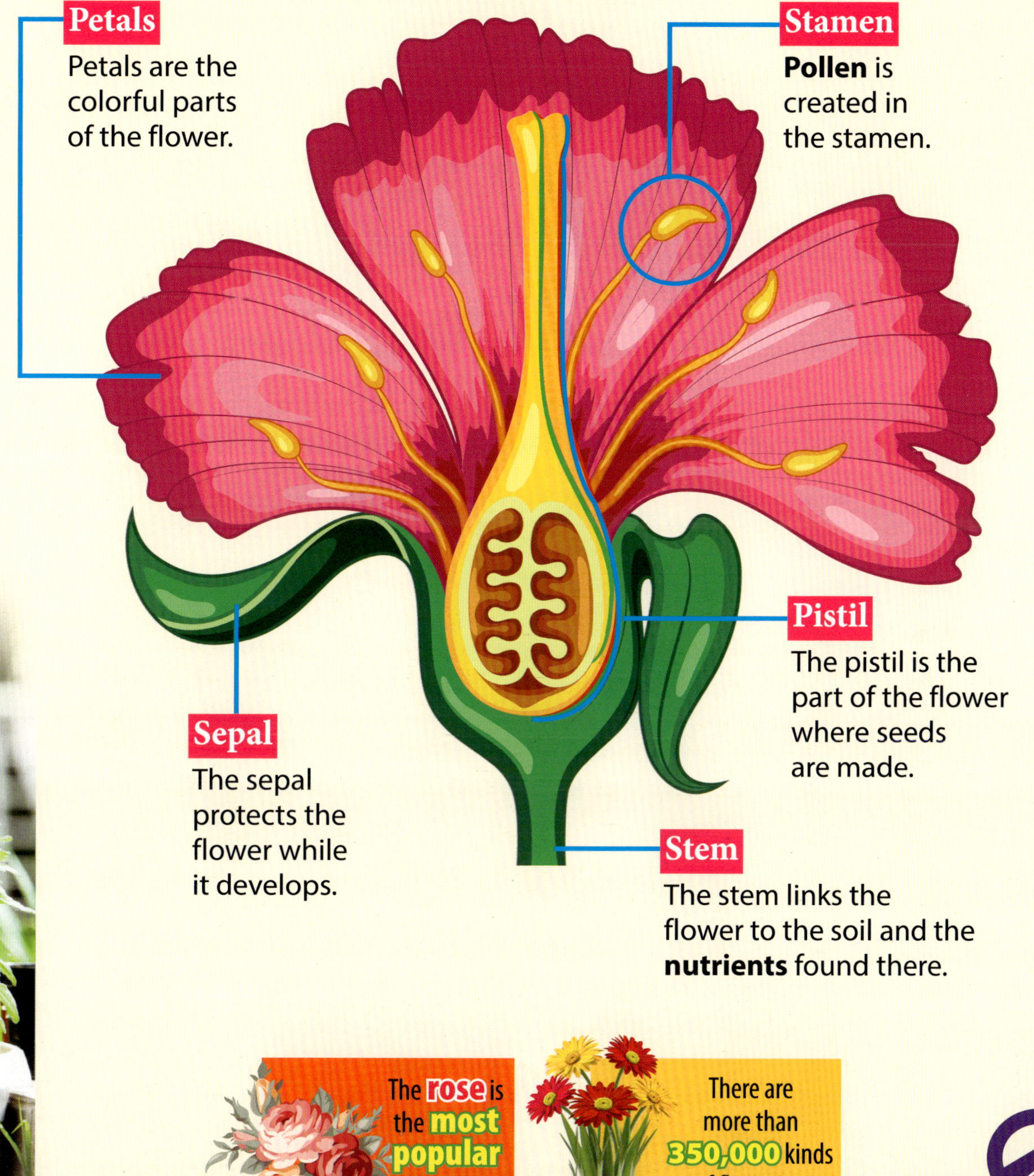

The **rose** is the **most popular flower** in the world.

There are more than **350,000** kinds of flowering plants.

Why Plant Flowers?

Flowers add much beauty to the places where they grow. They also fill the air with **fragrance**. Flowers are important for other reasons as well.

Insects are drawn to a flower's bright petals. When they land on the flower, pollen sticks to their bodies. The pollen falls off on the next flower the insects visit. This helps flowers make their seeds. These seeds are needed for new plants to grow.

Bees drink a sweet liquid called nectar from flowers. They turn the nectar into honey.

Some flowers, such as pansies, can be eaten.

Many flowers, including jasmine, are used to make perfumes.

The Life Cycle of a Flowering Plant

All flowering plants begin life, grow, and make more flowers. This is the life cycle of a flowering plant.

Seeds and Bulbs

Flowering plants can be grown from seeds or **bulbs**.

Pollination

Insects and other animals carry pollen from one flower to another. New seeds are made. The cycle begins again.

Roots

Roots are the first parts of a plant to grow. They hold the plant in the ground.

Seedling

The stem is the next part to grow. When it sprouts from the ground, the plant is called a seedling.

Flower

The seedling forms leaves, and its flowers bloom.

When to Plant

Is winter cold where you live, or is it hot all year? Flowers can only grow in certain conditions. You have to make sure that you plant your flowers at the right time of year. They may not grow if you plant them when it is too cold.

Spring is a popular time to plant flowers. Some flowers, such as marigolds, can be planted in early spring. Others, such as begonias, should be planted after the last **frost**. Flowers that grow from bulbs are often planted in the fall. These flowers will be some of the first to bloom in the spring.

SPRING

Spring is when most planting takes place.

SUMMER

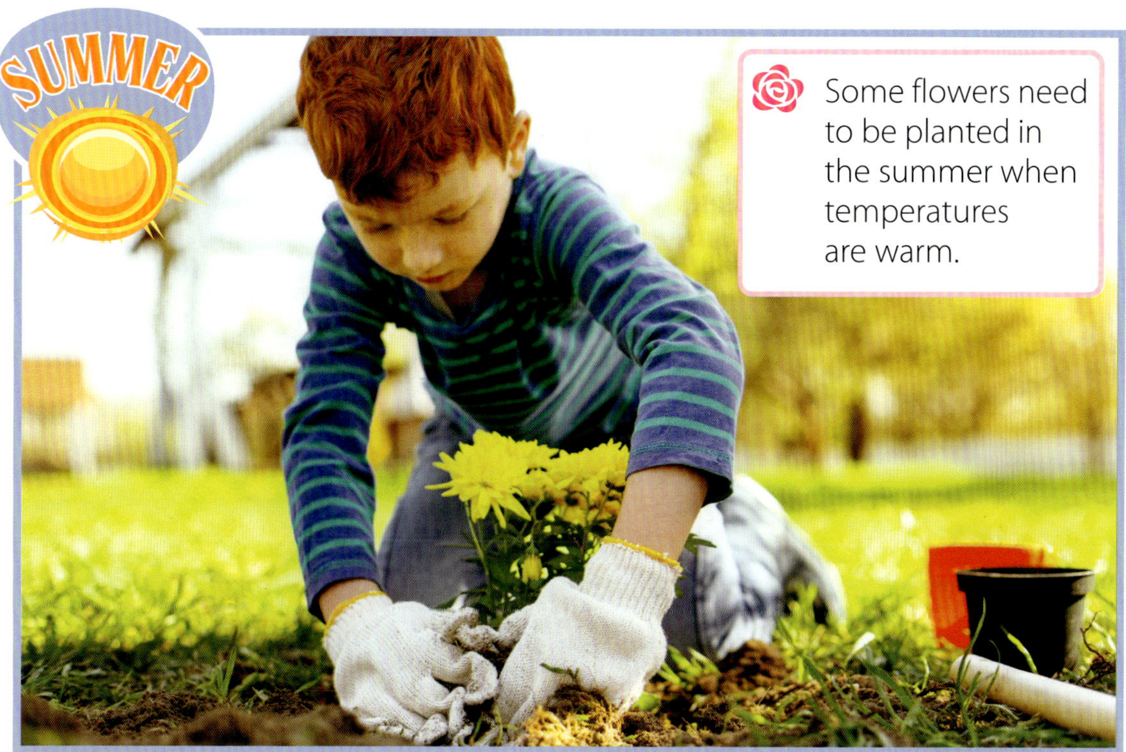

Some flowers need to be planted in the summer when temperatures are warm.

FALL

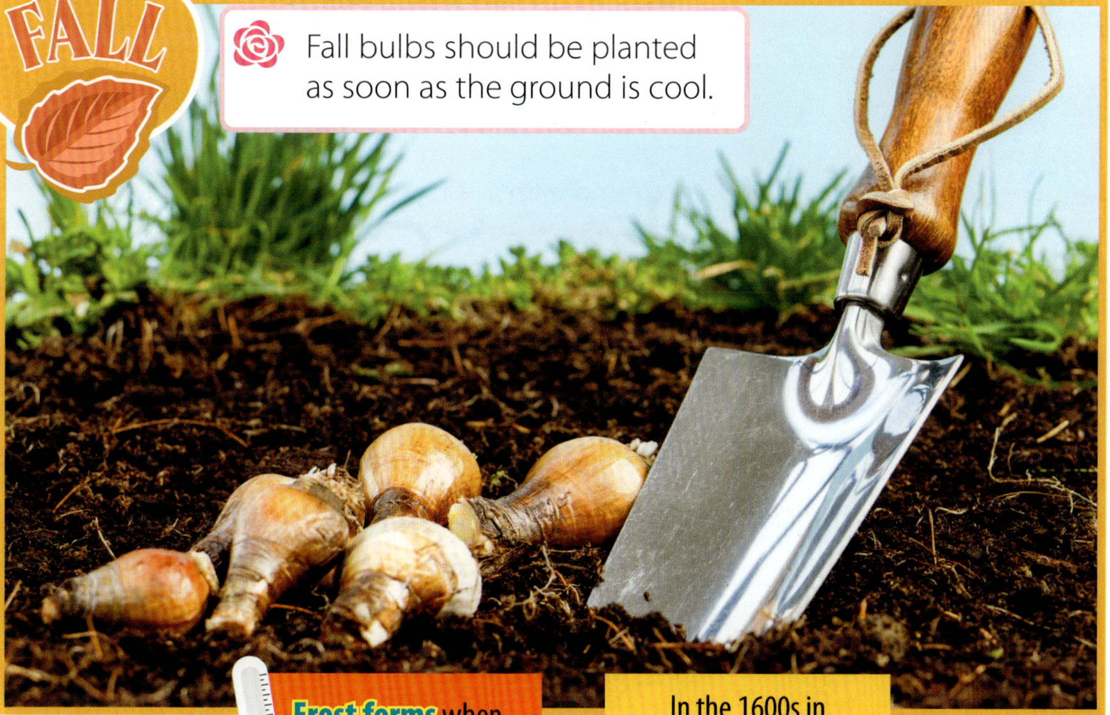

Fall bulbs should be planted as soon as the ground is cool.

Frost forms when the temperature **dips to 32°** Fahrenheit (0° Celsius).

In the 1600s in **Holland**, **tulip bulbs** were **worth more than gold**.

Where to Plant

It is important that you choose the right place to plant your flowers. How do you know if a place is right? There are three main factors you should consider.

Sunlight

Your garden should be somewhere that gets sunlight. Remember that some flowers like to be in shade for part of the day. Others prefer to spend all day in the Sun.

Water

All plants need water to survive. Place your flowers in an area where water can reach them. If rain does not fall, water your flowers with a watering can or a sprinkler.

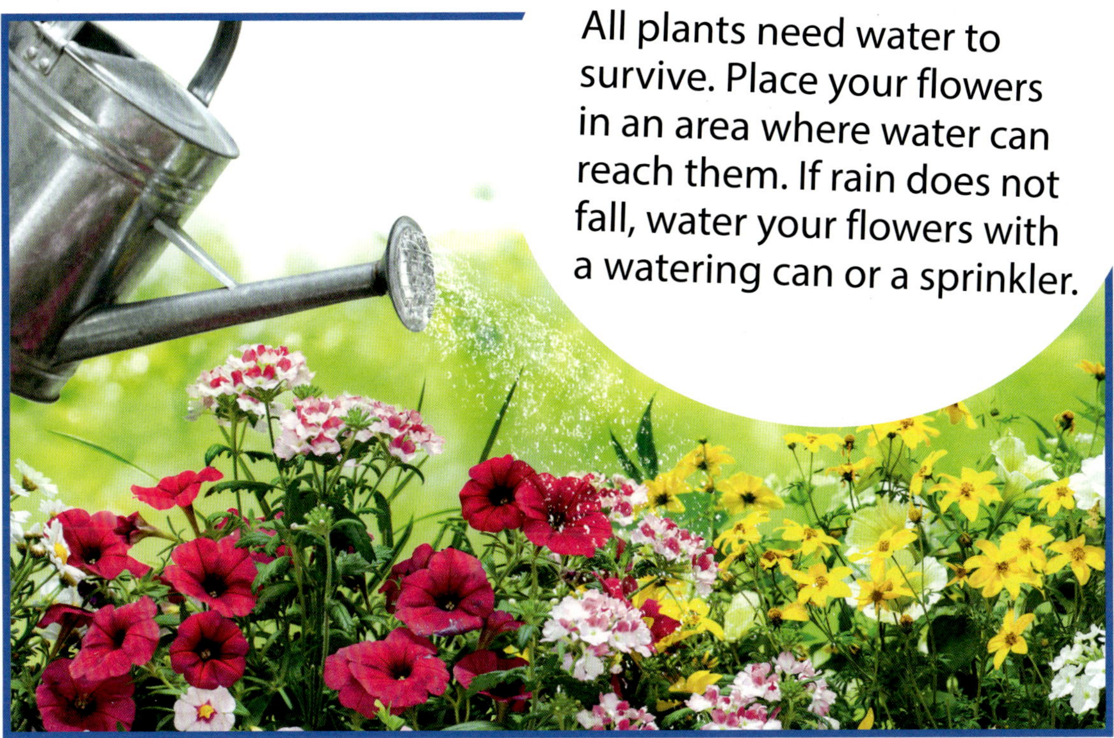

Nutrients

Make sure your flowers are planted in the right soil. Flowers need 17 different nutrients to help them survive. Fourteen of these nutrients come from soil.

Flower Gardens in the United States

Alaska

SCALE
0 500 miles
0 500 kilometers

Hawai'i

SCALE
0 100 miles
0 100 km

Washington
Montana
Oregon
Idaho
Wyoming
Nevada
Utah
Colorado
California
Arizona
New Mexico

1

N
W
E
S

LEGEND

- California
- Texas
- Missouri
- New York
- Other Land

SCALE

0 200 miles
0 200 km

1 San Francisco Botanical Garden

San Francisco, California

This garden is home to more than 100 **species** of rare magnolias.

2 Dallas Arboretum and Botanical Garden

Dallas, Texas

Visitors come to this garden every spring to see more than 500,000 flowers in bloom.

The United States has many well-known flower gardens. They are called **botanical** gardens because a wide variety of plants can be found there. Many of these gardens are open to the public. People are welcome to visit them and take in all of the colorful blooms.

3

Missouri Botanical Garden

St. Louis, Missouri

This 79-acre (32-hectare) garden is known for its large collection of rare orchids.

4

Brooklyn Botanic Garden

Brooklyn, New York

Founded in 1910, the Brooklyn Botanic Garden has more than 18,000 kinds of plants.

Choosing Your Flowers

There are so many flowers in the world. How do you decide which ones you should plant in your garden? One of the first things you have to think about is where you live. You may love one type of flower, but it might not be able to grow in certain **climates**. Also think about how much time you have to take care of your garden. Caring for some kinds of flowers takes more time than others.

Color is an important feature of a garden. Most gardeners choose plants that complement each other in color.

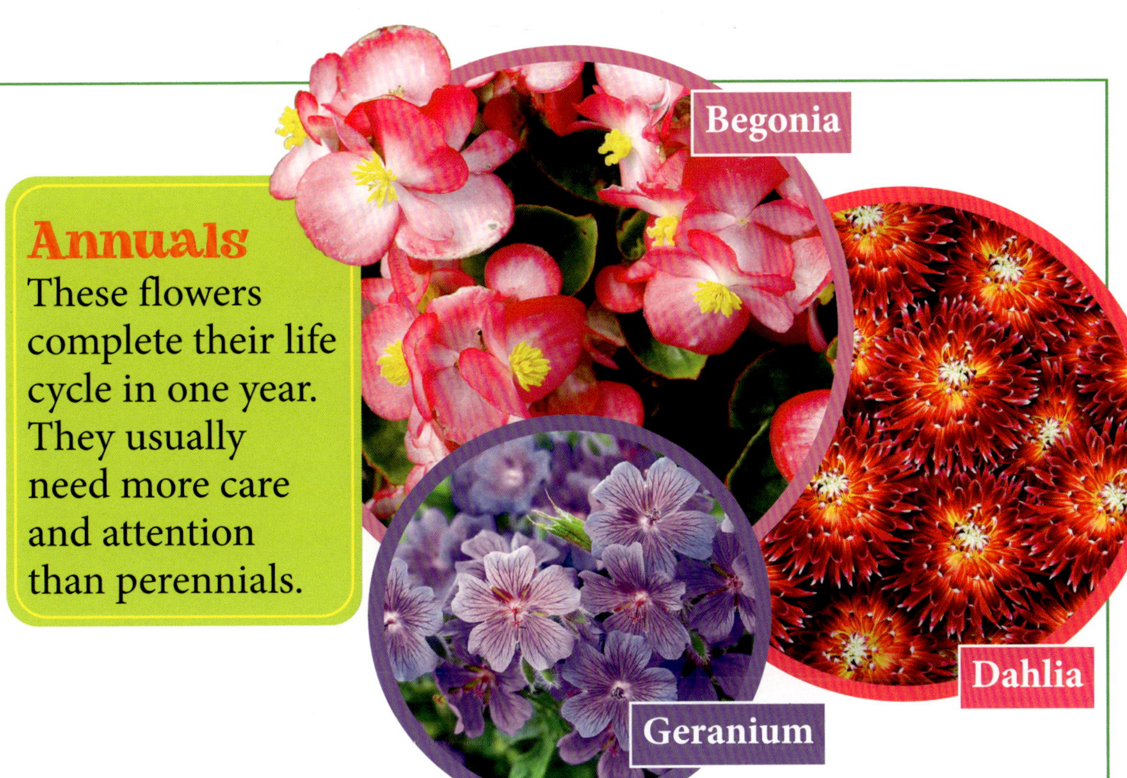

Annuals

These flowers complete their life cycle in one year. They usually need more care and attention than perennials.

Begonia
Dahlia
Geranium

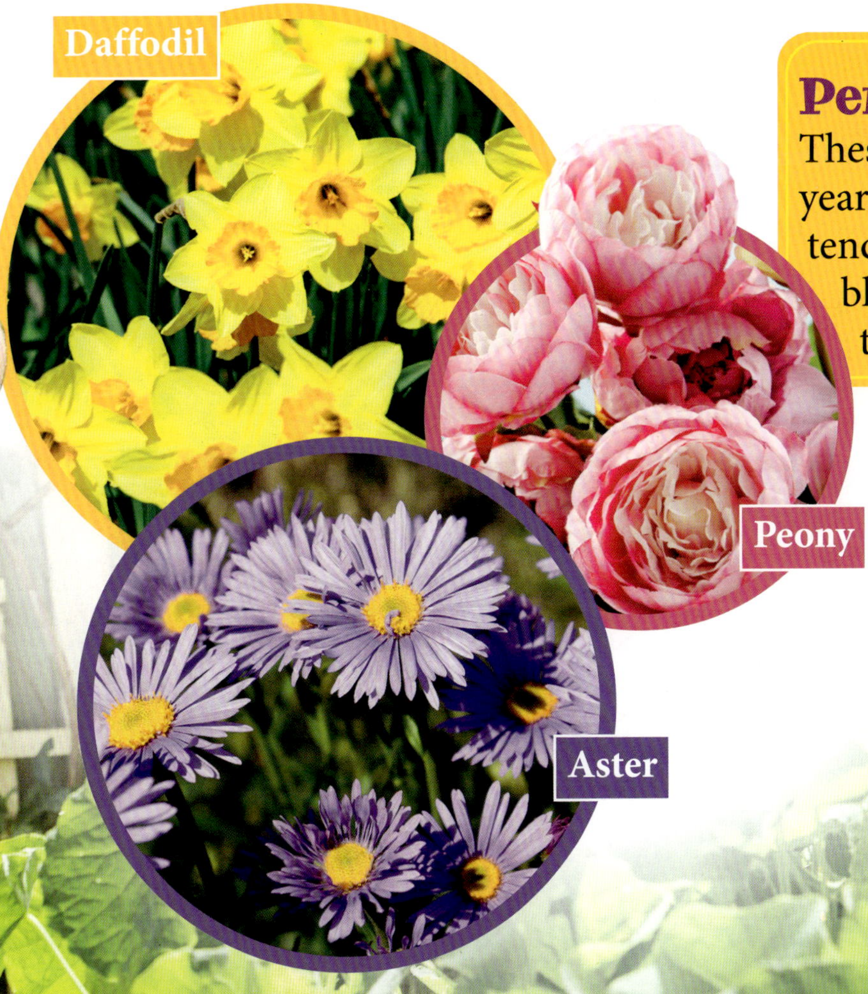

Daffodil
Peony
Aster

Perennials

These flowers bloom year after year. They tend to have shorter blooming periods than annuals.

Ready to Plant

If you want to grow a successful garden, you have to start with the right equipment. You also have to follow the planting process.

Gardening Equipment

Shovel

A shovel has many uses in gardening. It is used to loosen soil, dig holes, and move dirt out of the way.

Trowel

A trowel is like a small shovel. It is used to place flowers in the ground and to dig up weeds.

Watering Can

A watering can is used to deliver water to specific plants or to provide a sprinkling of water over a small area.

The Planting Process

1

Use your shovel to loosen the soil.

2

Take your trowel and dig a small hole for the flower seed. The hole's depth should be twice the width of the seed.

3

Place the seed in the hole.

4

Put the soil back in the hole and gently pack it down with your hands.

5

Use your watering can to soak the soil around the flower seed.

Repeat for each flower you want to plant.

Flower Garden Care

Taking care of a flower garden does not stop with planting. You must check on it every day to make sure that the plants are growing and staying healthy. There are three main areas that could need attention.

Watering

Check that the soil is not too dry. It is best to give your flowers a good soaking once or twice a week.

Plant Food

Spraying plant food on your flowers will help them grow. The food has the nutrients your flowers need.

Weeds

Weeds take up the space, nutrients, and water your flowers need to grow. Make sure to remove any weeds that you see.

10 Question Flower Quiz

1 Which flower is the most popular in the world?

2 Into what two groups can flowers be placed?

3 Which group of flowers completes its life cycle in one year?

4 What are the first parts of a plant to grow?

5 What process lets flowering plants make new seeds?

6 True or false? Some flowers like to spend part of the day in the shade.

7 When was the Brooklyn Botanic Garden founded?

8 What is a trowel used for?

9 How many times a week do most flowers need to be watered?

10 Why should weeds be removed from flower gardens?

ANSWERS 1. The rose 2. Annuals and perennials 3. Annuals 4. The roots 5. Pollination 6. True 7. 1910 8. To place flowers in the ground and to dig up weeds 9. Once or twice a week 10. They take up the space, nutrients, and water flowers need to grow.

Key Words

bloom: the colorful head on a mature flower

botanical: relating to plants

bulbs: a round, underground part of some plants from which the plants grow

climates: the prevailing weather conditions of a region

fragrance: a pleasant, sweet smell

frost: a thin coating of ice that forms when the temperature falls below freezing

nutrients: substances that promote growth

pollen: a dust made by plants that helps them develop seeds

species: groups of living organisms that share common features

Index

Get the best of both worlds.

AV2 bridges the gap between print and digital.

The expandable resources toolbar enables quick access to content including **videos**, **audio**, **activities**, **weblinks**, **slideshows**, **quizzes**, and **key words**.

Animated videos make static images come alive.

Resource icons on each page help readers to further **explore key concepts**.

Published by AV2
276 5th Avenue
Suite 704 #917
New York, NY 10001
Website: www.av2books.com

Library of Congress Cataloging-in-Publication Data
Names: Smith, Ryan and Kissock, Heather, authors.
Title: Flowers / Ryan Smith and Heather Kissock.
Description: New York, NY : AV2, [2022] | Series: Gardening | Includes index. | Audience: Grades 2-3
Identifiers: LCCN 2020022329 (print) | LCCN 2020022330 (ebook) | ISBN 9781791127718 (library binding) | ISBN 9781791127725 (paperback) | ISBN 9781791127732 (ebook other) | ISBN 9781791127749 (ebook other)
Subjects: LCSH: Flowers--Juvenile literature.
Classification: LCC SB406.5 .S65 2021 (print) | LCC SB406.5 (ebook) | DDC 635.9--dc23
LC record available at https://lccn.loc.gov/2020022329
LC ebook record available at https://lccn.loc.gov/2020022330

Printed in Guangzhou, China
1 2 3 4 5 6 7 8 9 0 25 24 23 22 21

022021
101120

Project Coordinator Heather Kissock
Designer Terry Paulhus

Photo Credits
Every reasonable effort has been made to trace ownership and to obtain permission to reprint copyright material. The publishers would be pleased to have any errors or omissions brought to their attention so that they may be corrected in subsequent printings.
AV2 acknowledges Getty Images, Minden Pictures, iStock, and Shutterstock as its primary photo suppliers for this title.